ENERGY SECTOR STANDARD OF THE PEOPLE'S REPUBLIC OF CHINA

中华人民共和国能源行业标准

Technical Code for Real-Time Water Temperature Monitoring Systems of Hydropower Projects

水电工程水温实时监测系统技术规范

NB/T 10386-2020

Chief Development Department: China Renewable Energy Engineering Institute

Approval Department: National Energy Administration of the People's Republic of China

Implementation Date: February 1, 2021

China Water & Power Press

中国水利水电出版社

Beijing 2024

All rights reserved. No part of this publication may be reproduced, stored in a retrieval system, or transmitted in any form or by any means—electronic, mechanical, photocopying, recording or otherwise, without prior written permission of the publisher.

图书在版编目（CIP）数据

水电工程水温实时监测系统技术规范：NB/T 10386-2020 = Technical Code for Real-Time Water Temperature Monitoring Systems of Hydropower Projects（NB/T 10386-2020）：英文 / 国家能源局发布. -- 北京：中国水利水电出版社，2024. 7. ISBN 978-7-5226-2601-7

Ⅰ. TV72-65

中国国家版本馆CIP数据核字第202424M76A号

ENERGY SECTOR STANDARD
OF THE PEOPLE'S REPUBLIC OF CHINA
中华人民共和国能源行业标准

Technical Code for Real-Time Water Temperature
Monitoring Systems of Hydropower Projects
水电工程水温实时监测系统技术规范
NB/T 10386-2020
（英文版）

Issued by National Energy Administration of the People's Republic of China
国家能源局　发布
Translation organized by China Renewable Energy Engineering Institute
水电水利规划设计总院　组织翻译
Published by China Water & Power Press
中国水利水电出版社　出版发行
　　Tel: (+ 86 10) 68545888　68545874
　　sales@mwr.gov.cn
　　Account name: China Water & Power Press
　　Address: No.1, Yuyuantan Nanlu, Haidian District, Beijing 100038, China
　　http: //www.waterpub.com.cn
中国水利水电出版社微机排版中心　排版
北京中献拓方科技发展有限公司　印刷
184mm×260mm　16 开本　2.25 印张　71 千字
2024 年 7 月第 1 版　2024 年 7 月第 1 次印刷
Price（定价）：￥360.00

Introduction

This English version is one of China's energy sector standard series in English. Its translation was organized by China Renewable Energy Engineering Institute authorized by National Energy Administration of the People's Republic of China in compliance with relevant procedures and stipulations. This English version was issued by National Energy Administration of the People's Republic of China in Announcement [2023] No. 8 dated December 28, 2023.

This version was translated from the Chinese Standard NB/T 10386-2020, *Technical Code for Real-Time Water Temperature Monitoring Systems of Hydropower Projects*, published by China Water & Power Press. The copyright is reserved by National Energy Administration of the People's Republic of China. In the event of any discrepancy in the implementation, the Chinese version shall prevail.

Many thanks go to the staff from the relevant standard development organizations and those who have provided generous assistance in the translation and review process.

For further improvement of the English version, any comments and suggestions are welcome and should be addressed to:

China Renewable Energy Engineering Institute
No. 2 Beixiaojie, Liupukang, Xicheng District, Beijing 100120, China
Website: www.creei.cn

Translating organizations:

POWERCHINA Zhongnan Engineering Corporation Limited

China Renewable Energy Engineering Institute

Translating staff:

GUO Yulan   LIU Xiaofen   LI Qian   ZHAO Kun

ZHANG Kankan   YE Yuxin   YUAN Yuan

Review panel members:

QIAO Peng	POWERCHINA Northwest Engineering Corporation Limited
YU Weiqi	China Renewable Energy Engineering Institute
PENG Qidong	China Institute of Water Resources and Hydropower Research

ZHU Jianzhong	Hohai University
WANG Yuanming	Sichuan University
DONG Haiying	POWERCHINA Kunming Engineering Corporation Limited
GAO Yan	POWERCHINA Beijing Engineering Corporation Limited
ZHANG Qian	POWERCHINA Guiyang Engineering Corporation Limited
LI Shisheng	China Renewable Energy Engineering Institute

National Energy Administration of the People's Republic of China

翻译出版说明

本译本为国家能源局委托水电水利规划设计总院按照有关程序和规定，统一组织翻译的能源行业标准英文版系列译本之一。2023年12月28日，国家能源局以2023年第8号公告予以公布。

本译本是根据中国水利水电出版社出版的《水电工程水温实时监测系统技术规范》NB/T 10386—2020翻译的，著作权归国家能源局所有。在使用过程中，如出现异议，以中文版为准。

本译本在翻译和审核过程中，本标准编制单位及编制组有关成员给予了积极协助。

为不断提高本译本的质量，欢迎使用者提出意见和建议，并反馈给水电水利规划设计总院。

地址：北京市西城区六铺炕北小街2号
邮编：100120
网址：www.creei.cn

本译本翻译单位：中国电建集团中南勘测设计研究院有限公司
　　　　　　　　　水电水利规划设计总院

本译本翻译人员：郭玉兰　刘小芬　李　倩　赵　坤
　　　　　　　　　张侃侃　叶雨欣　袁　嫄

本译本审核人员：

　乔　鹏　中国电建集团西北勘测设计研究院有限公司
　喻卫奇　水电水利规划设计总院
　彭期冬　中国水利水电科学研究院
　祝建中　河海大学
　王远铭　四川大学
　董海英　中国电建集团昆明勘测设计研究院有限公司
　高　燕　中国电建集团北京勘测设计研究院有限公司
　张　倩　中国电建集团贵阳勘测设计研究院有限公司
　李仕胜　水电水利规划设计总院

国家能源局

Announcement of National Energy Administration of the People's Republic of China [2020] No. 5

National Energy Administration of the People's Republic of China has approved and issued 502 energy sector standards including *Technical Code for Real-Time Ecological Flow Monitoring Systems of Hydropower Projects* (Attachment 1) and the English version of 35 energy sector standards including *Series Parameters for Horizontal Hydraulic Hoist (Cylinder)* (Attachment 2).

Attachments: 1. Directory of Sector Standards
2. Directory of English Version of Sector Standards

National Energy Administration of the People's Republic of China

October 23, 2020

Attachment 1:

Directory of Sector Standards

Serial number	Standard No.	Title	Replaced standard No.	Adopted international standard No.	Approval date	Implementation date
…						
2	NB/T 10386-2020	Technical Code for Real-Time Water Temperature Monitoring Systems of Hydropower Projects			2020-10-23	2021-02-01
…						

Foreword

According to the requirements of Document GNKJ [2016] No. 238 issued by National Energy Administration of the People's Republic of China, "Notice on Releasing the Development and Revision Plan of Energy Sector Standards in 2016", after extensive investigation and research, summary of practical experience, consultation of relevant advanced standards of China, and wide solicitation of opinions, the drafting group has prepared this code.

The main technical contents of this code include: general provisions, basic requirements, system design, system construction, and system operation management and maintenance.

National Energy Administration of the People's Republic of China is in charge of the administration of this code. China Renewable Energy Engineering Institute has proposed this code and is responsible for its routine management. Energy Sector Standardization Technical Committee on Hydropower Planning, Resettlement and Environmental Protection is responsible for the explanation of specific technical contents. Comments and suggestions in the implementation of this code should be addressed to:

  China Renewable Energy Engineering Institute
  No. 2 Beixiaojie, Liupukang, Xicheng District, Beijing 100120, China

Chief development organizations:

  China Renewable Energy Engineering Institute

  POWERCHINA Zhongnan Engineering Corporation Limited

Participating development organizations:

  POWERCHINA Guiyang Engineering Corporation Limited

  POWERCHINA Chengdu Engineering Corporation Limited

  China Three Gorges Corporation

  Yalong River Hydropower Development Company, Ltd.

  Wuling Power Corporation Limited

  Sichuan University

Chief drafting staff:

ZHAO Kun	ZHANG Kankan	XUE Lianfang	YU Weiqi
QIU Jinsheng	YAN Zhonglin	LU Bo	JIANG Hao

YANG Jie	XU Jingcao	CHEN Min	LI Jinghua
LI Kefeng	TUO Youcai	XU Hengjian	LI Xiang
CHU Kaifeng	ZHANG Dejian	ZHAO Xinchang	LIU Zhongxin
ZENG Youcong	YANG Jingbiao	YANG Ping	YAN Jianbo
LIU Huangcheng			

Review panel members:

WAN Wengong	RUI Jianliang	CHEN Guozhu	WANG Chunyun
CUI Lei	YANG Hongbin	CHEN Bangfu	LI Min
DAI Xiangrong	ZHANG Rong	JIANG Hong	SHI Jiayue
KOU Xiaomei	LIU Guihua	ZHANG Demin	LI Yingxi
WU Wenping	ZHAO Dihua	LI Shisheng	

Contents

1	**General Provisions**	1
2	**Basic Requirements**	2
3	**System Design**	3
3.1	General Requirements	3
3.2	Basic Data	3
3.3	Planning of Station Network	4
3.4	Technical Scheme for Monitoring	5
3.5	Communication Network	6
3.6	Power Supply, Overvoltage Protection and Earthing	7
3.7	Equipment Configuration of Telemetry Stations	8
3.8	Data Processing System of Central Station	10
3.9	Monitoring Facilities	12
3.10	Design Results	13
4	**System Construction**	14
4.1	Construction of Monitoring Facilities	14
4.2	Equipment Integration, Installation and Testing	14
4.3	Software Application and Management	14
4.4	Acceptance	15
5	**System Operation Management and Maintenance**	16
5.1	Management System	16
5.2	Maintenance of Facilities and Equipment	16
5.3	Data Management	16
Appendix A	Data Storage Codes of Water Temperature Monitoring	18
Appendix B	Contents of the Report on Design of Real-Time Water Temperature Monitoring System for Hydropower Project	19
Explanation of Wording in This Code		21

1 General Provisions

1.0.1 This code is formulated with a view to standardizing the scope, methods, criteria and regulations, and unifying the technical requirements for real-time water temperature monitoring systems of hydropower projects.

1.0.2 This code is applicable to the design, construction, operation management and maintenance of real-time water temperature monitoring systems of hydropower projects.

1.0.3 The real-time water temperature monitoring shall meet the general requirements of river basin eco-environment monitoring, and follow the principles of systematicness, representativeness and coordination.

1.0.4 In addition to this code, the real-time water temperature monitoring systems of hydropower projects shall comply with other current relevant standards of China.

2 Basic Requirements

2.0.1 The real-time water temperature monitoring system of a hydropower project shall be designed and constructed on the basis of the collection of adequate data on environmental conditions and project characteristics, taking into account the whole-process environmental management. System operation management regulations and procedures shall be formulated, and management and maintenance shall be performed.

2.0.2 The design and construction of real-time water temperature monitoring system of a hydropower project shall fully utilize the relevant monitoring conditions of the project, considering the accessibility, communication, management and maintenance, safety, etc. The real-time water temperature monitoring system of the project and those of the adjacent cascade hydropower projects shall be well coordinated and connected.

2.0.3 The planning of real-time water temperature monitoring system of a hydropower project shall take account of the layout of hydropower complex structures and selective withdrawal facilities, the reservoir operation mode, and the environment-sensitive objects, and shall be safe, reliable, technically feasible, and cost effective.

2.0.4 The real-time water temperature monitoring system of a hydropower project shall have the basic functions of automatically collecting, transmitting, storing, processing, displaying and sharing the water temperature data in real time.

3 System Design

3.1 General Requirements

3.1.1 The system design shall, based on the functional need analysis, determine the scheme of the real-time water temperature monitoring system through a techno-economic comparison.

3.1.2 The system shall be designed according to the water temperature monitoring requirements, taking account of the topographical and geological conditions, the project type, the layout of hydropower complex and selective withdrawal facilities, etc..

3.1.3 The design tasks of a real-time water temperature monitoring system shall include the following:

1 Define the basic functions of the system.

2 Propose the layout of station network.

3 Propose the technical scheme for monitoring.

4 Propose the design scheme of communication network.

5 Propose the power supply scheme as well as the overvoltage protection and earthing measures.

6 Propose the equipment configuration scheme of telemetry stations.

7 Propose the data processing system scheme of the central station.

8 Propose the construction scheme of monitoring facilities.

9 Propose the scheme of system operation management and maintenance.

10 Prepare the cost estimate of the system establishment.

3.2 Basic Data

3.2.1 The relevant data on environmental protection of the hydropower project shall be collected, including the documents of environmental impact assessment, post environmental impact assessment and environmental protection design.

3.2.2 The data on hydropower project characteristics shall be collected, including:

1 Characteristic parameters and regulation performance of the reservoir.

2 Characteristic parameters and design drawings of the hydropower complex.

3 Characteristic parameters and design drawings of the selective withdrawal facilities.

3.2.3 The data on the operation of the hydropower project shall be collected, including the reservoir operation scheme, reservoir inflow and outflow in typical years, water levels upstream and downstream of the dam, ice regime of the reservoir, and the operation data of selective withdrawal facilities.

3.2.4 The topographical data of the real-time water temperature monitoring sites with a survey accuracy of 1:2 000 or higher shall be collected.

3.2.5 The characteristic data of representative meteorological stations in the project area shall be collected, including air temperature, humidity, sunshine, wind speed, and wind direction.

3.2.6 The characteristic data of representative hydrologic stations in the project area shall be collected, including water level, flow rate, flow velocity, flood, and ice regime.

3.2.7 The relevant data on hydrologic monitoring systems and water-related structures in the project area shall be collected.

3.3 Planning of Station Network

3.3.1 The real-time water temperature monitoring station network shall be composed of the central station and telemetry stations. The planning of monitoring station network shall make full use of existing monitoring station networks, and may give priority to sharing the equipment, data and station rooms with relevant monitoring facilities.

3.3.2 The planning of real-time water temperature monitoring sections shall follow the principles of controllability and representativeness. Monitoring sections shall be set at proper places upstream and downstream of the dam. If necessary, additional monitoring sections shall be properly set downstream of the dam and at the head and middle of the reservoir, respectively.

3.3.3 The telemetry station sites shall meet the technical requirements of water temperature monitoring, be convenient for system communication networking, station construction and operation management, and avoid the areas where landslides, debris flow and other sudden disasters might occur as well as interference sources such as strong electromagnetic fields and strong vibration.

3.3.4 The equipment room of the central station shall be set up in a building with stable power supply, reliable network communication, good accessibility, good lightning protection and earthing measures, shall be away from factories, warehouses and storage yards with dust, smoke, harmful gases, or corrosive, flammable or explosive materials, and shall be free from interference of strong vibration source, strong noise source or strong electromagnetic field.

3.4　Technical Scheme for Monitoring

3.4.1　The technical scheme for real-time water temperature monitoring shall be developed according to the conditions of the telemetry stations and the water temperature monitoring requirements.

3.4.2　The layout of real-time water temperature monitoring sections shall meet the following requirements:

1. For the project with selective withdrawal facilities, a monitoring section shall be set upstream of the selective withdrawal facility near its centerline; for the project with a temperature control curtain, monitoring sections shall be set upstream and downstream of the curtain near its centerline, respectively.

2. For the power station at dam toe, a monitoring section shall be set at the junction of powerhouse tailwater and flood release outflow; for a conduit- or mixed-type hydropower station, monitoring sections shall be set below the dam and at the tailwater outlet, respectively.

3.4.3　The layout of vertical lines upstream of the dam for vertical water temperature monitoring shall meet the following requirements:

1. At least one vertical water temperature measuring line shall be arranged upstream of the dam, and additional measuring lines may be provided according to the transverse distribution of water temperature.

2. Vertical water temperature measuring lines should avoid the power intakes, flood release inlets and turbulent flow zones.

3. Vertical water temperature monitoring shall adopt floating or fixed temperature chains according to the topological conditions, hydrologic conditions, security performance, etc..

3.4.4　The layout of measuring points on vertical water temperature measuring lines shall meet the following requirements:

1. The measuring points on vertical water temperature measuring lines shall reflect the water temperature gradient along the water depth and follow the principle of arranging more measuring points in the upper water layer and less measuring points in the lower water layer. Additional measuring points shall be provided in the thermocline with a steep temperature gradient.

2　For freeze-up reservoirs, the measuring points shall be properly added on vertical water temperature measuring lines in the epilimnion according to the characteristics of ice regime.

3　The measuring points for reservoir outflow water temperature monitoring shall be set 0.5 m below the water surface or the underside of ice cover in the main stream zone of the river.

3.4.5 The frequency of water temperature monitoring should be once per hour.

3.4.6 For the vertical water temperature monitoring section upstream of the dam, water pressure sensors shall be set at the proper points of the temperature chain to measure the water depth synchronously.

3.5 Communication Network

3.5.1 The communication network shall ensure that the system can transmit monitoring data safely, reliably, accurately and timely, and perform data transmission between telemetry stations and the central station.

3.5.2 The design of communication network shall include communication mode selection, channel testing, system communication network scheme, work system, and technical indicators of communication channels.

3.5.3 The communication network shall meet the following requirements:

1　The system communication mode and network scheme shall be determined on the basis of analyzing the communication environment conditions of water temperature telemetry stations and comparing the techno-economic indicators of system construction and operation alternatives.

2　The water temperature telemetry stations shall adopt main and spare channels for data transmission.

3　Data transmission channels shall adopt the communication mode characterized by low bit error rate and high signal strength.

4　Communication network may adopt the communication mode such as digital mobile communication, satellite, ultrashort wave, Wi-Fi, or their combinations.

3.5.4 The work system shall meet the following requirements:

1　The work system shall be determined according to the system function requirements, communication mode, network scheme, management and maintenance capability, as well as cost effectiveness and ease of maintenance.

2 The work system may adopt the self-reporting system, polling-answer system, or their combination.

3.5.5 Technical indicators of communication channels shall meet the following requirements:

1 Priority shall be given to public communication network, established private network and other existing communication network. The communication mode shall be determined on the premise of ensuring data transmission rate and reliability, and spare channels shall be provided. The bit error rate of data transmission channel for main communication mode shall be in accordance with Table 3.5.5.

Table 3.5.5 Bit error rate of data transmission channel for main communication mode

Communication channel	Digital mobile communication	Satellite	Ultrashort wave
Bit error rate	$\leq 1 \times 10^{-6}$	$\leq 1 \times 10^{-6}$	$\leq 1 \times 10^{-4}$

2 Quality test for main and spare channels shall be conducted at water temperature monitoring stations. The success rate of data transmission between telemetry stations and the central station shall not be lower than 98 %. For ultrashort wave networks, the noise margin of communication circuits shall be higher than 10 dB.

3.6 Power Supply, Overvoltage Protection and Earthing

3.6.1 The design of system power supply shall meet the following requirements:

1 The DC power supply voltage should adopt 12 V. The voltage fluctuation range and ripple voltage shall meet the equipment requirements, and the voltage fluctuation range shall be controlled within +20 % to −15 %.

2 The AC power supply shall adopt single-phase power supply of 50 Hz and 220 V, and the voltage fluctuation range shall be controlled within ±10 %.

3 The central station shall adopt uninterruptible power supply. The capacity of uninterruptible power supply and its storage battery shall be determined according to the power consumption of equipment and the

reliability of local AC power supply, and shall keep the communication receiving server operating normally for at least 8 h.

4 The telemetry stations should use solar panels to float charge the storage batteries.

5 Storage batteries shall be of the maintenance-free type and the capacity shall meet the requirements of the equipment working for 45 consecutive days without sunlight.

3.6.2 Overvoltage protection and earthing measures shall meet the following requirements:

1 The integrated earthing should be adopted for the protective earthing and functional earthing of various equipment, the lightning protection earthing of buildings, and the earthing of power frequency AC power supply equipment.

2 The earth resistance of the central station and telemetry stations shall not exceed 5 Ω and 10 Ω, respectively.

3 The communication interfaces of remote terminal units shall adopt photoelectric isolation measures. A lightning protection device should be provided for the feeder before entering the communication terminal unit.

4 For AC power supply, isolation transformers, surge absorbers and AC voltage stabilizers should be provided.

3.7 Equipment Configuration of Telemetry Stations

3.7.1 Equipment configuration of telemetry stations shall include sensors, remote terminal units, communication equipment, and power supply system.

3.7.2 Equipment configuration design of telemetry stations shall select the equipment that can meet the corresponding technical requirements, through a comprehensive comparison in terms of cost, service life, type, etc. If necessary, backup equipment shall be provided.

3.7.3 Equipment configuration of telemetry stations shall meet the following requirements:

1 The technical indicators, parameters and number of main equipment in the telemetry station shall be determined according to the design results of system telemetry station network, communication and power supply.

2 The equipment shall be certified by the authorized quality inspection agencies.

- 3 The input, output and communication interfaces of sensors shall match those of the remote terminal units.

- 4 The equipment of telemetry stations shall operate normally in the site environmental conditions. Auxiliary means shall be taken to ensure the normal operation of equipment when the temperature at the site is out of the range of -40 °C to +85 °C, or the relative humidity is higher than 95 %.

3.7.4 Remote terminal units shall meet the following requirements:

- 1 The functions of remote terminal units shall include:

  1) Automatic data collection, storage, remote transmission and power supply management as well as expanding sensor interfaces and communication interfaces and upgrading software.

  2) Submitting information regularly.

  3) Setting parameters locally or remotely.

  4) Supporting sensors with interfaces of 4 mA to 20 mA, 0 V to 5 V, SDI-12, RS485/RS232, and 1-wire.

  5) Two or more communication interfaces and supporting various wireless communication channels.

  6) Conducting self-inspection and transmitting the operating status of remote terminal units.

  7) Manual setting and data sending.

  8) Solid-state storage of monitoring data for over 1 year, which can support the function of central station-commanded data measurement.

- 2 Technical indicators of remote terminal units shall meet the following requirements:

  1) The working voltage should adopt DC 12 V.

  2) The working current shall not exceed 100 mA.

  3) The static on-duty current shall not exceed 3 mA for the self-reporting type, and 15 mA for the polling-answer back type and the mixed type.

  4) The annual clock error shall be less than 2.5 min.

  5) The mean time between failures (MTBF) shall be at least 25000 h.

3.7.5 Main technical indicators of sensors shall meet the following requirements:

1　Water temperature sensors shall meet the following requirements:

   1) The measuring range is -5 °C to 35 °C.

   2) The resolution is 0.01 °C or better.

   3) The maximum permissible error is 0.05 °C.

   4) The mean time between failures (MTBF) shall be at least 25000 h.

2　Water pressure sensors shall meet the following requirements:

   1) The resolution is 1.0 cm or better.

   2) The maximum permissible error is 3 cm when the water level fluctuation is 10 m or below, and 0.3 % when the water level fluctuation is over 10 m.

   3) The mean time between failures (MTBF) shall be at least 25000 h.

3.7.6 Communication terminals shall meet the following requirements:

1　The interfaces shall meet the technical requirements of communication.

2　The working voltage should be DC 12 V.

3　The mean time between failures (MTBF) shall be at least 25000 h.

3.8　Data Processing System of Central Station

3.8.1 The central station shall be equipped with communication equipment, computers, network devices and power supply system.

3.8.2 The data processing system of the central station shall be designed according to the requirements of the water temperature monitoring system. It includes the functional software of data receiving, storing, querying, warning, etc., the system software, and the system hardware equipment.

3.8.3 Data receiving, storing, querying, warning and other functions of the system shall meet the following requirements:

1　The system shall be capable of automatically processing, checking and decoding data messages.

2　The system shall have the functions of automatically storing data to the database and making backup. The data storage codes of water temperature monitoring shall be in accordance with Appendix A of this code.

3   Data querying results should be presented by chart and the GIS-based method may be adopted for querying.

4   Short messages and WeChat messages may be sent when there are warnings about anomaly of communication devices, data overlimit and abnormal working conditions of telemetry stations by self-checking of the central station.

5   The system shall reserve information exchange interfaces.

3.8.4 The hardware equipment configuration of the central station shall meet the following requirements:

1   The equipment shall meet the functional requirements of data processing system.

2   Computer network security protection and isolation measures shall meet the safety protection requirements.

3   The technical performance of equipment shall meet the following requirements:

   1) Data receiving devices shall be determined according to the communication mode of the telemetry stations.

   2) Data processing servers should adopt highly reliable industrial-purpose computers or commercial servers.

   3) The memory, hard disk capacity, clock frequency and information exchange rate of data processing servers shall meet the requirements of online storage data volume and system data processing response speed.

   4) Data processing devices shall have backups.

4   Equipment configuration shall meet the following requirements:

   1) The data processing system shall be subjected to reliability design and should be equipped with independent database servers and disaster recovery backup database servers. Isolation devices, firewalls and other security protection devices shall be equipped when it is connected to other information systems.

   2) Data receiving shall be subjected to process design to determine data receiving devices, backup devices and UPS.

   3) Computer network shall be subjected to logical structure design to determine workstations, servers, switches, routers, etc.

4) Peripheral equipment, such as printers, projectors, large screen displays, and video surveillance devices of equipment room, shall be determined according to the system application requirements.

5) The working environment conditions of central station equipment shall meet the requirements that the temperature is between 18 °C and 28 °C and the relative humidity is between 40 % and 70 %. If the requirements are not met, auxiliary means shall be taken to ensure the normal operation of equipment.

3.8.5 The software configuration of the central station shall meet the following requirements:

1. Data processing software shall include system software and application software.

2. System software shall meet the system functional requirements, including the software of operation system, database management system, basic office, common programming, anti virus, and anti-illegal access. The names, versions and number of system software shall be specified.

3. Application software shall include the software function modules of monitoring station management, communication management, data receiving and processing, database storage, real-time monitoring and warning, remotely commanded data measurement, remote setting, data editing and querying, data compilation and reporting, data transmission and exchange, information dissemination, etc.

3.9 Monitoring Facilities

3.9.1 The design of monitoring facilities shall put forward the technical indicators and requirements of the monitoring facilities of the system, and work out the items and scales of monitoring facilities.

3.9.2 The design of monitoring facilities shall take account of the local conditions, make full use of existing facilities, and meet the requirements of equipment normal operation and system maintenance and management.

3.9.3 The telemetry station should be an integrated structure, and shall be provided with station foundation, safety fences, access roads, signal boards, etc.

3.9.4 The wind-resistant design of the telemetry stations shall be based on local meteorological conditions. The basic wind pressure shall be converted from the 10 min mean maximum wind speed with 15 years return period.

3.9.5 The machine room of the central station shall have lightning protection, fire prevention and anti-static functions.

3.10 Design Results

3.10.1 The design of the real-time water temperature monitoring system of a hydropower project shall present a design document, a list of monitoring technological requirements, a bill of quantities, a list of equipment, a list of data storage code, tables of cost estimate, a general layout of monitoring system, a system plan, typical sections, a data transmission topology diagram, a management system diagram, etc.

3.10.2 The contents of the report on design of real-time water temperature monitoring system for hydropower project should be in accordance with Appendix B of this code.

4 System Construction

4.1 Construction of Monitoring Facilities

4.1.1 Prior to construction of monitoring facilities, field reconnaissance shall be conducted, and a construction plan shall be developed.

4.1.2 The construction of monitoring facilities shall be carried out by a professional construction team after site preparation.

4.1.3 Upon completion of the construction of monitoring facilities, acceptance shall be conducted.

4.2 Equipment Integration, Installation and Testing

4.2.1 Equipment integration and installation shall be carried out after the monitoring facilities and equipment are accepted.

4.2.2 Relevant tests and simulation operations shall be carried out after the equipment integration.

4.2.3 Equipment checks shall include mechanical check and electrical performance check. Tests and on-line debugging shall be conducted if necessary.

4.2.4 Equipment installation shall be carried out in accordance with the procedures specified in the user manuals or instructions, and fastening and waterproof treatment shall be well done. The parameter settings of equipment shall be checked and the equipment performance shall be tested to ensure its normal operation.

4.2.5 The process of equipment installation and testing shall be recorded, and a report on system installation and testing shall be prepared.

4.3 Software Application and Management

4.3.1 Application software shall follow the principles of standardization and normalization. The software design, software coding, protocol development and documentation shall comply with relevant regulations.

4.3.2 The software shall be capable of realizing the functions specified in the design scheme.

4.3.3 The software shall adopt the object-oriented program design and fault-tolerant design, and be subjected to general-purpose packing, encapsulation and interface calling.

4.3.4 The software shall be characterized by high reliability, data security, convenient operation, ease of maintenance, and expandability.

4.4 Acceptance

4.4.1 A trial run shall be carried out for at least 3 months after the installation and testing of system equipment.

4.4.2 After the trial run, the project owner shall organize the system acceptance.

4.4.3 Acceptance materials shall include the system design report, contract documents, system installation and testing report, system construction report, system trial run report, etc.

4.4.4 System acceptance shall assess the system functions, technical performance and conformance. Suggestions on system operation management shall be proposed.

5 System Operation Management and Maintenance

5.1 Management System

5.1.1 Regulations and procedures for operation management and maintenance of the monitoring system shall be formulated.

5.1.2 Skilled professionals shall be assigned to perform system operation management and maintenance.

5.2 Maintenance of Facilities and Equipment

5.2.1 The maintenance of monitoring facilities and equipment shall meet the following requirements:

1. The staff on duty shall monitor the operating status of the system, and take measures as soon as possible in the case of any problems.

2. The personnel shall be assigned to care the telemetry station.

3. The monitoring system shall be inspected regularly. The performance of facilities and equipment in telemetry stations shall be overall checked and tested to eliminate the malfunctions and replace the faulty parts.

4. The water temperature monitoring instruments shall be calibrated at least once in each hydrological year.

5. Routine maintenance shall be conducted to keep the equipment room clean and orderly, and ensure the system working environment conditions.

6. Unscheduled inspections shall be performed according to the operating status of the monitoring system.

7. The operation and maintenance of monitoring facilities and equipment shall be recorded.

5.2.2 When the overall performance of the system is significantly degraded under normal maintenance, the equipment shall be replaced by category or an overhaul shall be carried out.

5.3 Data Management

5.3.1 The archive on system construction and operation shall be established for data management, including the basic conditions of telemetry stations and the central station, equipment performance, maintenance records of equipment and facilities, and contact information of management personnel.

5.3.2 The system operation management and maintenance personnel shall compile the water temperature monitoring data every year, and prepare the annual report on system operation.

5.3.3 In accordance with the archive management regulations, the system operation management and maintenance personnel shall organize and archive the overall system design report, system installation and testing report, system construction report, system trial run report, acceptance documents, annual report on water temperature monitoring, calibration records of instruments and equipment, records of operation and maintenance, as well as related contracts, terms of reference, and written appraisal opinions.

Appendix A Data Storage Codes of Water Temperature Monitoring

Table A Data storage codes of water temperature monitoring

No.	Field name	Identifier	Type & length	Null value	Unit	Primary key index No.
1	Station code	STCD	C(8)	N	–	1
2	Time	TM	TIME	N	–	2
3	Monitoring point code	TPCD	VC(4)	N	–	3
4	Water depth	WD	N(5,2)	–	m	–
5	Water temperature	WT	N(4,2)	–	℃	–

Appendix B Contents of the Report on Design of Real-Time Water Temperature Monitoring System for Hydropower Project

Foreword

1 Introduction

2 Project Overview

2.1　River Basin Overview

2.2　Project Description

2.3　Project Operation and Scheduling

2.4　Selective Withdrawal Facilities

2.5　Environment-Sensitive Objects

3 Planning of Station Network

3.1　Planning Range

3.2　Planning Principle

3.3　Layout Scheme

4 Technical Scheme for Monitoring

4.1　Overall Scheme

4.2　Design of Monitoring Sections

4.3　Design of Vertical Monitoring Lines

5 Communication Network

5.1　Selection of Communication Mode

5.2　Selection of Work System

5.3　Design of Communication Network

6 Power Supply, Overvoltage Protection and Earthing

6.1　System Power Supply

6.2　Overvoltage Protection and Earthing

7 Equipment Configuration

7.1　Sensors

7.2　Telemetry Station Equipment

7.3   Central Station Equipment

8   Data Processing System of Central Station

8.1   Data Processing

8.2   Software Configuration

9   Monitoring Facilities

9.1   Central Station

9.2   Telemetry Stations

10   System Construction

11   System Operation Management and Maintenance

12   Cost Estimate

Explanation of Wording in This Code

1 Words used for different degrees of strictness are explained as follows in order to mark the differences in executing the requirements in this code.

   1) Words denoting a very strict or mandatory requirement:

      "Must" is used for affirmation; "must not" for negation.

   2) Words denoting a strict requirement under normal conditions:

      "Shall" is used for affirmation; "shall not" for negation.

   3) Words denoting a permission of a slight choice or an indication of the most suitable choice when conditions permit:

      "Should" is used for affirmation; "should not" for negation.

   4) "May" is used to express the option available, sometimes with the conditional permit.

2 "Shall meet the requirements of..." or "shall comply with..." is used in this code to indicate that it is necessary to comply with the requirements stipulated in other relative standards and codes.